TAKE NOBODY'S WORD FOR IT

TAKE NOBODY'S WORD FOR IT

BOOK OF EXPERIMENTS

George Auckland and Bill Coates MBE

BBC BOOKS

Published to accompany a series of
television programmes in consultation with the
BBC Continuing Education Advisory Council

Published by BBC Books,
a division of BBC Enterprises Limited,
Woodlands, 80 Wood Lane, London W12 0TT
First Published 1989
© The authors 1989
Reprinted 1989 (six times), 1990, 1992 (three times)

ISBN 0 563 21447 3
Cover photograph by Chris Capstick
Back cover photos by Hendrick Ball
Cover design by Pep Reiff
Illustrations by Kay Dixey

Set in 11/14pt Linotron Bodoni by
Phoenix Photosetting, Chatham, Kent
Printed and bound in Great Britain by
Mackays of Chatham PLC, Chatham, Kent
Cover printed by Fletchers of Norwich

For Zoë, who first put her finger into the looking-glass world before she was two

CONTENTS

INTRODUCTION

Much of our knowledge of the world is second-hand – we've read it in a book or paper, heard it on the radio or seen it on tv. This book, which accompanies our series *Take Nobody's Word For It*, is about a different world where direct or 'hands-on' experience is every bit as valuable as words.

Many museums now have sections where you can 'play' with the exhibits. If the word 'play' sounds strange think about the words of Nobel prize-winning physicist Erwin Schrödinger: 'Science develops from the play of a civilised society.' So, play.

The experiments shown in our programmes were very popular so I hope this book contains enough information for you to try them out successfully. If you can't get some of them to work always remember they didn't work first time for us either. It's quite possible you'll discover more if the experiments don't work immediately.

As far as possible, we've avoided using equipment which scientists are supposed to use. You'll find no test-tubes here – yoghurt pots rule, OK? Should any experiment not work because of a mistake in the text then I apologise – you can't even take our word for it!

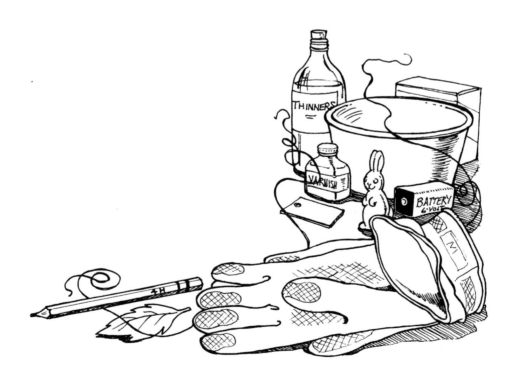

PRINT YOUR OWN T-SHIRT LOGOS

Why wear other people's designs when you can wear your own?

You will need:

A washed and ironed cotton T-shirt

A Xerox or Nashua photocopy of your design

A wallpaper edging roller

White spirit and water in a jam jar

Washing-up liquid

A warm metal tray (heated with an iron)

A paint brush

What to do:

Mix together two parts water, one part white spirit and a good squirt of washing-up liquid in a jam jar. Paint the mixture on the back of your photocopy. Turn the paper over and paint some mixture on the picture. Be more careful with this side. Leave to soak.

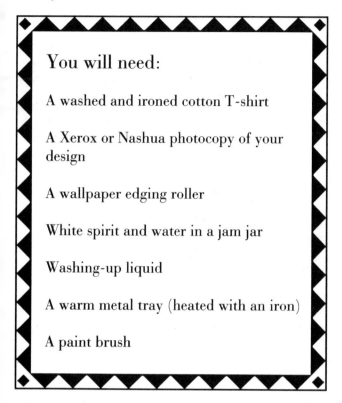

Put the warm metal tray inside the T-shirt and straighten out the material. Then drain the photocopy and place it face down in position on the T-shirt. Use the wallpaper roller to press the photocopy picture to the material. Work carefully to avoid creasing the paper, but use as much firm pressure as you can. Make sure all parts of the picture have been well rolled. Peel away the photocopy and you should have a good transfer on the T-shirt. It's a good idea to try this out first on an old T-shirt or piece of cloth, because not all brands of photocopy work – and you may forget that pictures transferred in this way are reversed!

To make your print washproof, you need to coat it with a waterproofing mixture (from sports shops) or fixative spray (from art shops).

The experiment depends on mixing together two liquids which normally will not mix: white spirit and water. The washing-up liquid has long molecules which attract oily stuff at one end and water at the other, and this enables the spirit and water to mix and form an emulsion. (The same thing happens in your washing-up bowl.) The emulsion mixes with the carbon in the photocopy to form an ink which transfers to the T-shirt.

EXOTIC MULTICOLOURED CARNATIONS

If you stand a carnation in a pot of ink and water, it will slowly turn the same colour as the ink. But this experiment gives you much more interesting effects.

You will need:

White carnations or irises

Small yoghurt pots

Coloured inks and food dyes

What to do:

Mix together ink, food dye and water. The exact amounts depend on the colours used, but for a start try adding 5 teaspoons of ink and 5 drops of food dye to a standard-size yoghurt pot filled with water.

Trim the lips off two yoghurt pots so they stand very close together. Mix up a red solution in one pot and blue in the other, stirring well. Shorten the stem of a white carnation and split it lengthwise.

Put one half of the split stem into each yoghurt pot, so one half is dipped in red and the other half in the blue solution (you may have to support the flower).

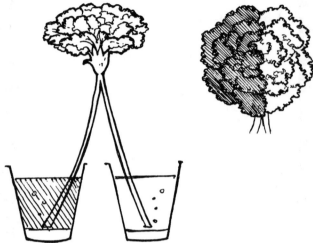

After a few hours in a warm room, the carnation will become half-red and half-blue as the inky water rises up tiny tubes inside the stem and colours the petals.

You can vary the effect by using different amounts of ink and food dye – the red mixture need not be as strong as blue or green.

If you want to colour an iris, try splitting the stem three or four ways and place it in different colours. The effect can be really exotic, with each petal turning a different colour. The mixture of ink and food dye produces a much better colour than just ink on its own.

The experiment shows that once water enters the tiny tubes in the stem of a flower, it never mixes again but is delivered to specific parts of the flower.

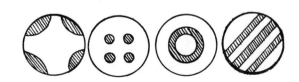

THE MAGDEBURG TUMBLERS

Recreate a famous experiment first performed by Otto von Guericke in Magdeburg, Germany, in 1654.

What to do:

Put a small piece of aluminium foil at the bottom of one tumbler to prevent the plastic getting scorched. Soak a piece of blotting paper in water, light a match and drop it on to the foil in the tumbler. Put the blotting paper on top of the tumbler to make a seal and place the other tumbler upside-down on top of the blotting paper. Make sure the lips of the two tumblers are well lined up. Press them lightly together.

When the match has gone out, pick up the top tumbler. The bottom one should come with it because you've created a partial vacuum inside. Outside atmospheric pressure keeps them together.

The burning match heats up the air which expands out of the bottom tumbler before you put on the wet

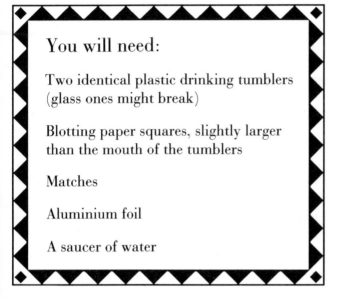

In his experiment von Guericke joined together two hollow hemispheres, pumped out the air and then tried to get two teams of horses to drag them apart. They failed and so showed what a powerful effect the atmosphere can have as the hemispheres were only being held together by atmospheric pressure.

paper. The air inside then cools and the pressure drops. Also the match uses up the oxygen which combines with the hydrocarbons in the wood to make water. The water takes up less space than the original oxygen, and this reduces the pressure too.

THE MYSTERIOUS RATTLEBACK

A spinning-top which changes direction by itself.

You will need:

A plastic teaspoon

A piece of cardboard 1½ × 3in (4 × 8cm) *or* a piece of clear plastic ruler (better) *or* a pencil

Modelling clay

A smooth, hard surface

Now set the strip at an angle to the long axis of the spoon. Spin it again. If you have set the angle about right, the rattleback will start to behave rather strangely. It will spin, then wobble; it will then stop spinning and begin to spin in the opposite direction. Try different angles between the spoon and the strip until you get this reversal.

What to do:

Cut or break the bowl from the plastic teaspoon, making sure that the bowl doesn't crack. Fill the bowl carefully right up to the top with modelling clay. Place the cardboard strip, ruler or pencil centrally on the modelling clay, along the long axis of the spoon. Spin the object on a smooth hard surface (such as a mirror tile). It will behave as you would expect. Try spinning it in the opposite direction too.

The reasons why the rattleback changes direction are complicated. However, they depend on the bowl of the spoon having a long and short radius of curvature and the long axis being out of alignment with the cardboard strip.

A PARTY TRICK FOR EGGHEADS

*How do you tell a raw egg from a hard-boiled egg —
without breaking them?*

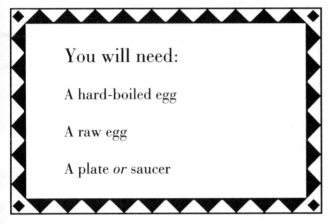

You will need:

A hard-boiled egg

A raw egg

A plate *or* saucer

Do the same with the other egg. Again, note what happens to it. One of the eggs should start to spin again.

The contents of this egg are liquid and carry on swirling even when the shell has temporarily stopped. When you take your finger away the swirling viscous contents transfer some of their energy to the shell and the egg spins again. So, the respinning egg is raw.

Eggsactly what you would eggspect!

What to do:

Place one egg on the plate and spin it. Touch it with your finger until it stops and then let go quickly (easier to do than describe!). Note what happens to the egg.

EGGSITING EGGSPERIMENTS (Yuk!)

If you've ever had problems beating egg white you'll know it's peculiar stuff. For a start it's one of the few foods which is an alkali (test it with a little of the red cabbage water, page 27). Here are a few ideas to help you improve, or worsen, your performance.

You will need:

Egg white (be careful to exclude any yolk)

A copper bowl if you can get one

Cream of tartar

A glass bowl

Cooking oil

What to do:

Beat egg whites in glass and copper bowls and compare the stability of the foam.

Add about $\frac{1}{16}$ teaspoon of cream of tartar to an egg white and beat it in a glass bowl. Copper bowls and cream of tartar should help to produce more stable foam.

Add a small amount of cooking oil to an egg white and beat it. The oil noticeably reduces the amount of foam and also makes it collapse. Egg yolks, which are very fatty, have a similar effect.

Cream of tartar is acidic and reduces the alkalinity of the egg white. The acid probably affects the protein molecules.

MOVING AIR DEFIES COMMON SENSE

Hang two beach balls from strings so that they are a few inches apart. With a hair drier blow air through the gap between them. What will happen? You may be surprised.

You will need:

Two beach balls *or* two ping-pong balls

Two lengths of string *or* cotton

A hair drier *or* a drinking straw

Blu-Tack, a pencil and bottles

What to do:

You can do this experiment on a large scale with beach balls or a small scale with ping-pong balls. Attach short lengths of cotton to the ping-pong balls with small pieces of Blu-Tack. Tie the cotton to a pencil and balance the pencil between two bottles.

As you blow the air through the gap, instead of blowing apart as you might expect, the balls touch together. You'll have to adjust the separation of the balls to give the best effect – it depends how hard you blow!

This is called the Bernoulli Effect. The air which is moving between the balls has a lower pressure than the surrounding stationary air, so the balls are pushed together.

There are other simple experiments that demonstrate this effect. Try cutting a strip of paper about 1½ × 6 in (4 × 15 cm). Hold one end on your bottom lip and blow gently. The paper will rise up to near horizontal.

Cut a strip of paper and fold it to the shape shown, like a bridge. Stand it on the edge of a table. Blow gently through the bridge and see what happens.

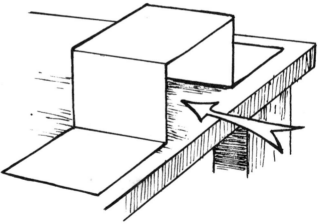

The Bernoulli Effect is very important because it helps to explain how aeroplanes can fly. The shape of the wing is important. Air goes over the top and underneath the wing. As the wing moves, the air has to travel a greater distance over the top than underneath, so it moves at a higher speed and the pressure drops. (The higher the speed the lower the pressure.) Now the pressure under the wing is greater than above it and this makes the wing lift and the plane fly.

MORE EXPERIMENTS WITH MOVING AIR

You will need:

A hair drier

A ping-pong ball

A bottle

A candle

Try tilting the hair drier. The ball will stay inside the column of air even when the drier is strongly tilted.

Put two ping-pong balls in the vertical column. Sometimes you can get them to dance around each other.

PS If you happen to tread on the ping-pong ball in the course of the experiment (as I did) you can get the dent out by putting the ball in a mug and pouring boiling water over it.

What to do:

Place a ping-pong ball in the stream of air blowing vertically upwards from the hair dryer. The ball will stay within the column of air. Watch what happens to it?

Blowing through glass

Light a candle and stand a bottle in front of it. Blow hard on the opposite side of the bottle – the candle will flicker and go out.

But don't take my word for it.

GLUTEN FROM FLOUR

Cooks, parents and doctors often talk about gluten, but how many of you have seen it and handled it?

You will need:

Flour (bread flour is best)

A bowl

Water

What to do:

Make a dough from 2 oz (50 g) of flour and 2 table-spoonsful (30 ml) of water. Knead it until it's smooth. Soak the ball of dough in water for at least half an hour, leaving overnight if possible.

Gently squeeze and knead the dough under a trickling tap until the water dripping from the dough is no longer milky. The milky water contains starch and what's left is gluten – it's a bit like chewing gum. Gluten consists of long intertwined molecules which give it elasticity.

When you are sure all the starch is washed away, bake the gluten ball in a hot oven, gas mark 8, 450°F (230°C) for about 20 minutes. You should end up with a perfect miniature loaf. In the oven, the water has turned into steam and puffed out the gluten into a light open structure. This is the framework or skeleton for an ordinary loaf of bread.

Let the starch settle out of the water and keep it for another experiment (see page 33).

THE DIY ELECTROSTATIC GENERATOR

If you rub an ordinary vinyl LP with a piece of wool or nylon it will attract lots of dust. The rubbing causes the record to become charged with static electricity.

You will need:

An old 12-inch record

Lots of other things too numerous to mention (see diagram)

What to do:

The diagram is self-explanatory. The main thing to watch out for is good insulation. At high voltages many materials such as wood and glass conduct electricity (because of moisture). Modern plastic bottles are fantastically good insulators. The generator should easily produce half inch long sparks.

It will drive many electrostatic toys such as this simple motor. Balance a rotor blade on a pin stuck in plasticine and stand it on the left-hand door knob. Electrons flying off from the sharp points of the rotor kick it backwards (obeying Newton's third law of motion – *Action and reaction are equal and opposite*).

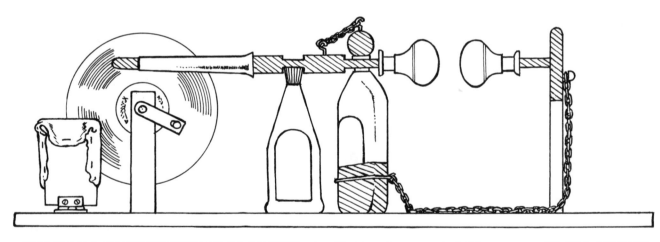

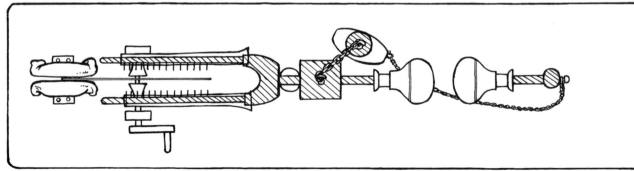

THE DIY ZOETROPE

The Zoetrope was invented in 1834 by William George Horner. The word means 'movement of life'.

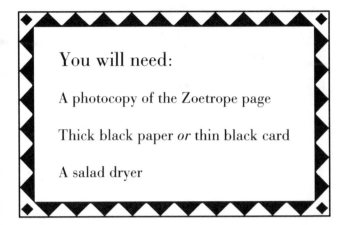

You will need:

A photocopy of the Zoetrope page

Thick black paper *or* thin black card

A salad dryer

What to do:

The top of a salad dryer is a very useful geared rotating platform which can be pressed into service for many experiments and still dry the salad.

Cut out the pieces of the zoetrope and glue them to the thick paper. Cut out a cardboard disc 6 in (15 cm) in diameter, to form the base. Cut out the slits from the zoetrope. Join the two sections together by gluing flap A under edge A and flap B under edge B to form a cylinder. Attach the cylinder to the cardboard base. Make a hole in the centre of the base so that it will fit on the top of the salad dryer or even your record player's turntable.

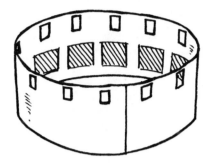

Spin the zoetrope and look through the slits to see a moving picture.

If you have a 6 in (15 cm) diameter baking tin, you can fix the zoetrope to the inside, making sure you can still see through the slits. The baking tin can then be fixed to the salad dryer using pieces of Blu-Tack or sticky tape, thus making the zoetrope more stable.

The zoetrope works for the same reason that film and television pictures work – persistence of vision. If your eye is shown a bright picture, even for a very short time, it retains the image. So a succession of similar pictures can be blended into one continuous image.

You can use the layout of the zoetrope to make your own 12-frame cartoon strips. They work best if the movement is a continuous cycle, like the horse galloping or someone running.

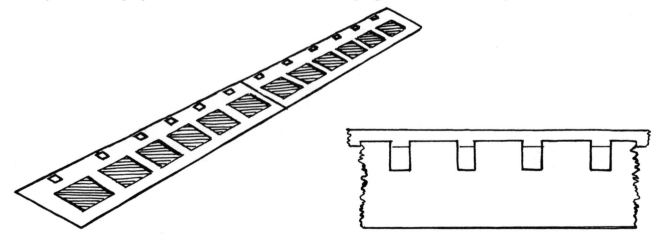

EDGE B   EDGE A

FLAP A   FLAP B

22

REVOLUTIONARY ILLUSIONS

Kitchen utensils in the service of science and fun.

You will need:

A salad dryer

Photocopies of Benham's Disc

The Zoetrope

What to do:

Cut out a photocopy of Benham's Disc and put it on the platform of the salad dryer. Turn the handle and look at the disc. At certain speeds and lighting conditions you should see distinct colours.

Fluorescent lighting usually produces a good effect. Spin the disc the other way.

If you don't have a salad dryer, glue the disc on to thick card, put a pencil through the centre and spin it like a top.

Why do we see colours when the discs are just black and white? Part of the answer might be that the red, green and blue colour receptors in our eyes have different persistence of vision. The different flickering effects from parts of the disc cause varying activity in the receptors and this produces the sensation of colour.

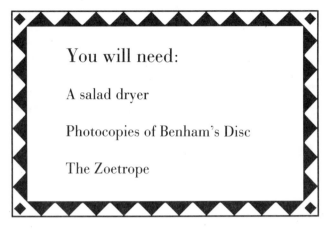

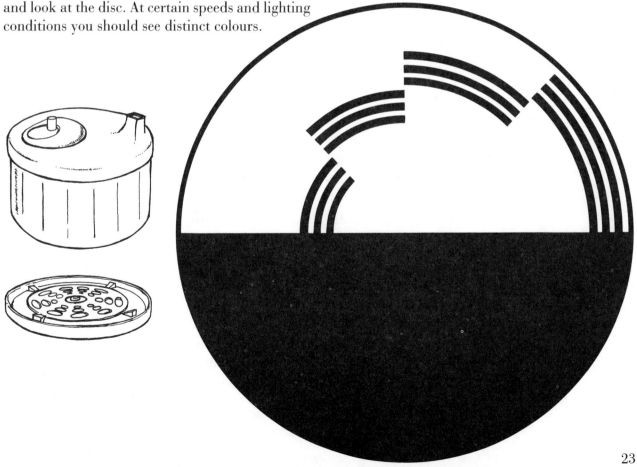

A CARTESIAN DIVER

This scientific toy is named after the seventeenth-century philosopher René Descartes – can anyone tell me why? The diver sinks or floats at your command.

You will need:

A plastic fizzy-drink bottle with its cap

A short length of transparent plastic drinking straw

Blu-Tack *or* similar stuff

A bowl

What to do:

Fill the bottle with cold water (cold boiled water is better).

To make the diver, cut a 3 in (8 cm) length from the straw and plug one end with Blu-Tack. Roll out a small length of Blu-Tack and wrap it round the other end of the straw. Put the diver into a bowl of water, it should float upright with the open end at the bottom. Adjust the amount of Blu-Tack until it floats with about 1 in (2.5 cm) of straw out of the water.

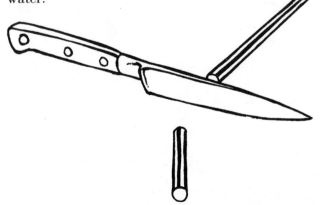

Tip the diver over and by tapping the straw allow water to enter. The idea is to adjust the buoyancy of the diver until it only just floats. If the diver sinks, hold it vertically out of the water and tap the top end, this will allow a small amount of water to drop out of the straw.

When the diver is floating properly, transfer it to the bottle, taking care not to lose any water. Top up the bottle and put its cap on tight.

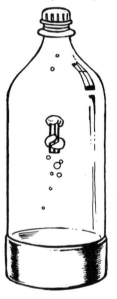

If you squeeze the bottle gently the diver should sink, release your hand and it should rise. As you squeeze, look closely at the straw, you should see the water-level inside it rise. Because the bottle is completely full of water, when you squeeze the only gas you can compress is inside the diver. So more water enters the bottom of the diver making it heavier until it sinks.

Fish and submarines alter their buoyancy by taking in water.

SMART COLOURS

Many of the coloured dyes around us are mixtures of several pigments. This experiment separates them out.

You will need:

Some Smarties *or* coloured pens

White blotting paper *or* coffee filter paper

Surgical spirit
or salt solution (1 teaspoon of salt in a pint of water)

A jug *or* large yoghurt pot

What to do:

Cut some of the paper into strips about 1 in (2.5 cm) wide and several inches long. Fold over one end of a strip so that it will hook over the side of the jug or pot and the other end will nearly touch the bottom. Lick a Smartie and rub it on to the strip of paper about 1 in (2.5 cm) from the bottom end. Make sure plenty of the colour is transferred. Alternatively, make a mark with a coloured pen.

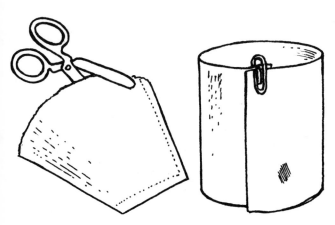

Pour surgical spirit or salt solution into the jug to a depth of about ½ in (1 cm). Hang the strip of paper over the edge of the jug so that the bottom dips into the liquid. Leave it for a while. Liquid will climb up the strip by capillary action and when it gets to the colour that too will be carried up.

In a short time you should see bands of various colours up the strip, especially if you started with something like brown, green or black. This is a technique called chromatography.

The choice of liquid is important since the dyes have to be soluble in it. Surgical spirit works well with Pentel pens. Salt solution might be better with water-based dyes.

If the experiment is done carefully you can work out which actual dyes are present in any mixture by comparing the results with known dyes.

A SIMPLE REACTION TIMER

This is a cheaper, but more accurate version of an experiment which usually uses a £5 note.

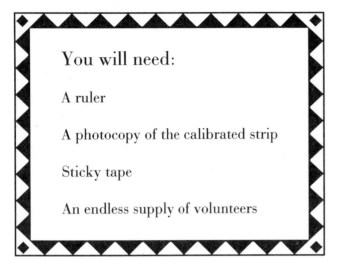

You will need:

A ruler

A photocopy of the calibrated strip

Sticky tape

An endless supply of volunteers

What to do:

In a famous experiment Galileo showed that all objects fall at the same rate, if we can ignore air resistance. We can use this observation to make an accurate reaction timer.

Trace out the calibrated strips and stick them along the flat side of a ruler. Get your volunteers to hold their forefinger and thumb either side of the ruler, just in line with the zero on the calibration, but not actually touching. You hold the ruler at the top end and tell the guinea pig (sorry – volunteer) to catch it as soon as it starts to fall.

Let the ruler drop without warning.

The reaction time can be read off directly from the scale at the point they gripped it between their thumb and finger.

If they are useless, you might need to extend the scale a bit. (The distances from the zero line for 0.22, 0.23, 0.24 and 0.25 seconds are 9½ in (23.7 cm), 10½ in (25.9 cm), 11 in (28.2 cm) and 12 in (30.6 cm) respectively.)

HOLD
HERE

SECS.
0.21
0.20
0.19
0.18
0.17
0.16

0.15
0.14
0.13
0.12
0.11
0.10
0.09
0.08
0.06
0.04
0.02

AN EXPERIMENT FOR CABBAGES AND KINGS
(and Prime Ministers)

THE ACID TEST

Red cabbage juice used as an indicator.

> You will need:
>
> Some red cabbage leaves
>
> Clear vinegar
>
> Bicarbonate of soda *or* washing soda *or* household ammonia

What to do:

Chop up some red cabbage leaves and simmer them in water for about 10 minutes.

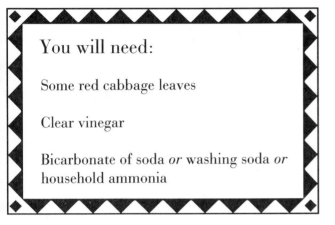

Pour off the liquid and dilute it with water until you have a rich purple liquid. Pour some of this liquid into three separate glasses. Add a little clear vinegar to the first glass and a strong solution of bicarbonate of soda to the second. Compare these colours with the third glass. You should have pink from the acidic vinegar and blue from the alkaline bicarbonate, using the cabbage juice as an indicator.

Washing soda and ammonia are stronger alkalis and will turn the purple water bright green. You can use the cabbage water to test other household substances – bleach, washing powder, egg white or lemon juice.

Acidity and alkalinity are measured on the pH scale. The neutral point is pH7 (pure water); vinegar is about pH3, bicarbonate of soda pH9, household ammonia pH13.

ANAMORPHIC PICTURES

Anamorphic pictures have been distorted in some way to disguise the image. Often you have to look at them from an odd angle or via a special mirror to make sense of the image. Usually they are made just for fun but sometimes they have contained the faces of forbidden heroes from the past!

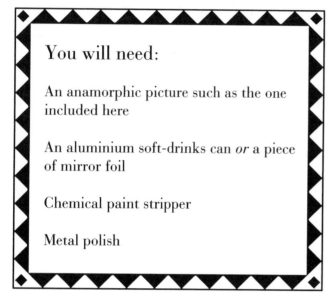

You will need:

An anamorphic picture such as the one included here

An aluminium soft-drinks can *or* a piece of mirror foil

Chemical paint stripper

Metal polish

The most common anamorphic pictures are viewed via a cylindrical mirror. These mirrors are not easy to obtain but the method below produces quite a good effect.

What to do:

Buy a soft drink (not cola) in an aluminium can (if you're not sure it's aluminium test it with a magnet – it shouldn't stick). Pick one which is free from dents and blemishes.

When it's empty strip away the paint using chemical paint stripper. *Follow the instructions which come with the chemical and be very careful. Wear gloves and avoid splashing it.* Some paint will strip more easily than others, trial and error will be needed until you find the best can to use.

Wash away all traces of the paint and stripper and dry the can. Polish the surface using metal polish or wadding-based polishing material. It will take a little time to get a good mirror-like surface and be careful to avoid buckling the can. Mirror foil (available from art shops) wrapped around a drinks can will do just as well.

When it's finished, place the mirror on the spot shown on the drawing and look at the reflection.

If you want to make an anamorphic drawing yourself, place the can in the centre of a blank sheet of paper and try drawing something simple like a square, then you'll see the kind of distortion needed. A set of grid lines will help with more complicated pictures.

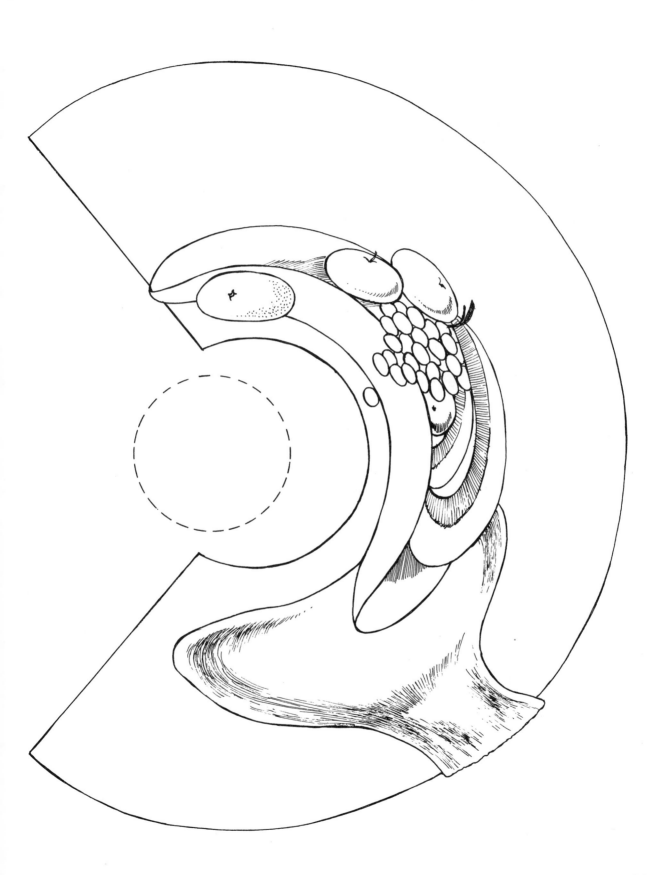

JAM JAR BALLET

How to make a gas do the work.

You will need:

A jam jar (the bigger the better)

Mothballs *or* raisins and small bits of dry pasta

Bicarbonate of soda

Vinegar (clear vinegar is best)

What to do:

Three-quarters fill the jar with tap water. Stir in as much bicarbonate of soda as will dissolve and then add some more. Pour in a few tablespoonsful of vinegar.

Drop in a few mothballs (handle carefully – they're poisonous) or some raisins and bits of pasta. They will sink to the bottom. Soon some of them will start to rise to the surface, then fall and rise again. The ballet has begun.

Look how the objects are covered in bubbles. This is carbon dioxide, formed when the acidic vinegar reacts with the sodium bicarbonate. All the objects you've thrown in the jar are denser than water and sink, but the bubbles of carbon dioxide which form on their rough edges give them lift and buoyancy.

If you use mothballs, look what happens when the balls break the surface of the liquid.

Should the ballet not work, add some more vinegar to the jam jar. Once it's started it will often perform for an hour or more. You may be able to find other small objects which will also work. They need to be slightly denser than water and have some rough edges.

A DIY BUBBLE CHAMBER

A simulated experiment from the world of nuclear physics.

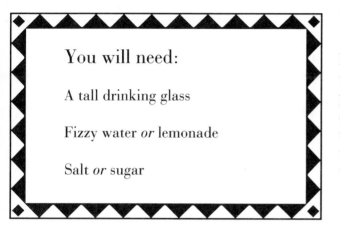

You will need:

A tall drinking glass

Fizzy water *or* lemonade

Salt *or* sugar

What to do:

Pour out a glass of fizzy water and leave it to stand until the first frothing is over. Sprinkle tiny heavy particles such as sugar, salt or sand into the drink and watch the trails of tiny bubbles which form as the particles sink. The rough corners of the particles allow the carbon dioxide, which is dissolved, to be released as a gas.

Bubble chambers are used by nuclear scientists to detect the paths of fast-moving sub-atomic particles such as protons and neutrons.

You may have noticed that when you release the cap from a new bottle of lemonade, beer or fizzy water, thousands of bubbles form in the liquid. This is carbon dioxide which is dissolved in the liquid at high pressure but becomes a gas again when the pressure drops.

In 1952 Dr Donald Glaser of Michigan University used this idea to make the first bubble chamber. He used a liquid called diethyl ether. The bubbles formed in the liquid when a high-energy particle had passed through and the pressure was suddenly dropped.

You can make a simple simulation of a bubble chamber on your kitchen table.

Bottles and glasses with good smooth inside surfaces will keep your fizzy drinks fizzier longer!

THE WORLD'S FIRST ELECTRIC MOTOR

In 1821, when Michael Faraday first made this electric motor, he used mercury. This is a safer version.

You will need:

An aluminium pie case

A piece of coat-hanger wire bent to shape

A lump of modelling clay

A piece of bare copper wire with a small hook bent at one end (the solid earth wire of a piece of mains cable is just the job)

A small strong cylindrical magnet (from a good tool shop)

A strong solution of salt in water (sea salt is a bit better)

A 6 volt battery and wires

Aluminium cooking foil

Pour some salt solution into the aluminium pie case. Connect the coat-hanger frame to the negative and the aluminium foil to the positive of the battery. The free end of the copper wire should start to swing around the magnet, and fizzing will occur round the copper wire. This is hydrogen being produced.

If nothing happens, try reversing the connections to the battery.

After the motor has been running for a few minutes, empty out the pie case and look at the bottom against a strong light. You'll see it's full of thousands of tiny holes where the aluminium has dissolved away. Any ideas for a use?

Michael Faraday first made his motor work one Christmas Day and this version came to Bill Coates also on Christmas Day in 1986, when his wife offered him a mince pie!

What to do:

Fix the coat-hanger wire into the lump of modelling clay. Put a small piece of paper underneath the clay to prevent the coat-hanger touching the foil. Assemble the motor as shown in the diagram – making sure the copper wire can swing freely around the magnet.

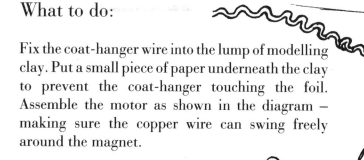

STARCH

THE BLACK AND WHITE TEST

Starch is present in many kinds of food. There is an easy test for it.

> You will need:
>
> Tincture of iodine (from a chemist – handle with care)
>
> Icing sugar
>
> Flour *or* cornflour
>
> Two saucers

What to do:

Place a small amount of icing sugar on one saucer and some cornflour on the other. Put one drop of iodine on each and see what happens. Iodine reacts with starch to produce a strong black colour.

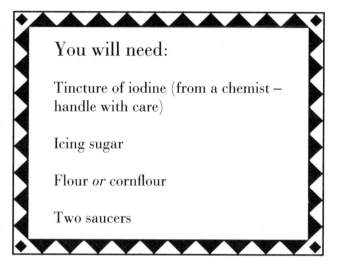

Which contains the starch, the icing sugar or the cornflour?

You can test for starch in many other things: potatoes, wallpaper paste, bread, plants and so on.

The ability of iodine to turn white starch black can be used in a fascinating experiment to produce a photograph on a geranium leaf (see page 36).

SWINGING SPUDS

What you might call a grandfarmer clock!

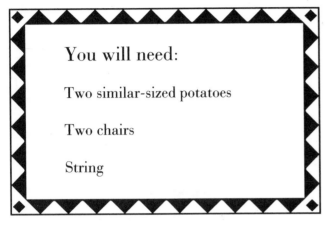

You will need:

Two similar-sized potatoes

Two chairs

String

What to do:

Stand the chairs back to back about 3 ft (1 m) apart and tie a length of string between them. Tie 18 in (46 cm) long pieces of string to two potatoes and fix these to the string between the chairs. The potatoes should be between 6 in (15 cm) and 1ft (30 cm) apart.

Start one potato swinging like a pendulum. After a few seconds the other potato will start to swing and eventually take over the motion. The first potato will stop dead – but not for long!

The swinging motion alternates between the potatoes because the strings are the same length and the potatoes are similar sizes. The system is said to resonate. Each time a potato swings its own string gives the horizontal string a little twist. This twist starts the other potato swinging. The twisting happens each swing at exactly the right time to give the other potato an extra push.

Try these variations:
Use two potatoes of widely different sizes
Have the strings different lengths
Suspend three or more potatoes

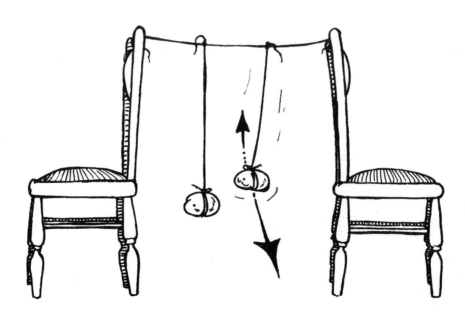

HOW TO RELIEVE THE TENSION

We think water is wet, but for many purposes such as washing it's just not wet enough. These experiments show just how dry water can be.

You will need:

A very clean bowl

Three matchsticks

Washing-up liquid

A sharp pencil

A medicine dropper *or* drinking straw

What to do:

Fill the bowl with water and float the matches in it. Bring them together to form a triangle. (Once in the shape they will stick together as if glued.)

Dip the pencil point in washing-up liquid and just touch the surface of the water inside the triangle. Kerpow! – the matches move apart as if exploding.

The surface of clean water acts as if it has a stretched dry skin. The effect is called surface tension. The same thing allows water to form droplets which are perfect spheres.

Use a medicine dropper to place a single drop of water on cloth such as wool or felt. It should hold together in a small ball and not wet the cloth. You can even remove it and leave the cloth completely dry.

Dip the pencil point in washing-up liquid and gently touch the ball of water resting on the cloth. The washing-up liquid reduces the surface tension and allows the water to wet the cloth.

Water molecules attract each other in all directions but the ones at the surface end up with spare unbalanced attractive forces. These forces cause the surface tension.

Soap and detergents can lower surface tension because they contain long molecules which attract water at one end and oily stuff at the other. They form a layer at the top of the water with the oil-loving ends of their molecules sticking upwards. These parts of the molecules don't attract each other as much as the molecules of plain water, so the surface tension is lower.

GERANIUM PHOTOGRAPHY

How to make geranium photographs by starving, exposing, boiling and otherwise mistreating an innocent geranium.

All green plants use light to help them produce food which they store. Geraniums keep the starch they make, in their leaves.

You will need:

A healthy geranium

A slide projector (if you can get one)

A slide of the photo you want to print – it should be a negative

Bicarbonate of soda

A cardboard box

A yoghurt pot

Methylated spirits

Tincture of iodine

Various jam jars

What to do:

Keep the geranium in a warm and completely dark place for about two days. By then it will have used up most of the starch stored in its leaves. Cut off a large leaf with plenty of stalk.

Make a hole in the cardboard box large enough for the stalk to go through. Once through the hole the stalk must dip into the yoghurt pot of bicarbonate

solution inside the box. Fold up a square of kitchen or loo paper, dip it in the bicarbonate, put it behind the leaf and fix the leaf to the box with small pieces of sticky tape. The leaf should now be vertical with its stalk dipped in the yoghurt pot.

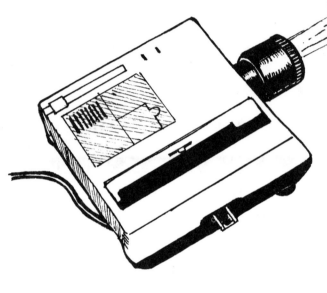

Line up the slide projector to produce a small sharp picture on the leaf (the lens of the projector will have to be well forward in its mount for this). The leaf needs a long exposure, depending on the brightness of the projector, but about 4 hours is a good start.

You can do this experiment without a slide projector. Just keep the geranium in the dark for 2 days as before. Then, without removing the leaf from the plant, clamp your negative to the leaf using two strips of thin glass and elastic bands. Leave it in daylight for a few hours.

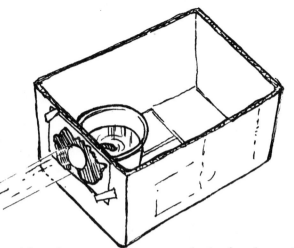

for 1 to 2 minutes to kill it. Next, drop it in a jam jar of methylated spirits. Stand the jam jar in hot water (careful with this bit!). Leave for about an hour.

The chlorophyll will dissolve in the meths and you will end up with a rather delicate, flabby and pale leaf. Wash it and place it in a bowl of warm water making sure it spreads out well. Pour in some tincture of iodine and shake the bowl gently. The picture should develop in about 30 seconds.

Amaze your friends.

After the exposure, remove the leaf and you should be able to see a faint but clear image. To develop the photo you first have to get rid of the green chlorophyll in the leaf. Dip the leaf in boiling water

The leaf-photo will have to be kept in water. (If it fades add more iodine.)

STATIC ELECTRICITY

There is only one kind of electricity. Whether it moves or not electricity is the result of electrons. The simplest way to obtain electrons is just by rubbing. They can produce some startling effects.

Scientific Balloons

An electric charge can be negative or positive. (Too many electrons or too few.) Do like-charges behave in the same way as the identical poles of magnets?

You will need:

Two round balloons

Cotton thread

What to do:

Blow up the balloons and tie each of them to the ends of 3 ft (90 cm) of cotton thread. Hold the thread in the middle and let the balloons hang down so they touch.

Now rub your hair all over the balloons. Each will take up the same charge from the rubbing action. Let them hang down again. If they attract each other, like-charges attract; if they repel each other, like-charges repel. They should repel!

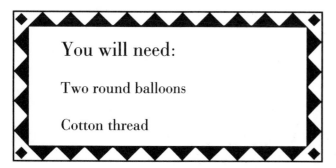

You can remove the charge by rubbing all over the surface of the balloons with the palm of your hand.

Comb and Water

You will need:

Water trickling from a tap

A plastic comb

What to do:

Turn on the tap and adjust it to produce the thinnest trickle of water that doesn't break into drops. Comb your hair then hold the teeth of the comb close to the water but without touching. On a good day with the right comb and hair you can get the water to turn through a right angle.

Opposite charges attract, like-charges repel. The water is attracted to the comb so it must have an opposite charge. Where does this charge come from?

As the water leaves the metal tap the water will have no charge, but water molecules themselves have positive and negative 'ends' and the charge on the comb makes them all line up.

THE DANCE IN THE FERRERO ROCHER BOX

Lots of modern plastic materials are very good electrical insulators and excellent for static electricity experiments.

You will need:

An empty Ferrero-Rocher sweetbox

Soft tissue such as loo paper *or* Kleenex

A plastic comb

A piece of felt *or* wool

Sticky tape

What to do:

Split the tissue into separate sheets and cut into strips 2 in (5 cm) long and ¼ in (0.5 cm) wide. Stick one end of some of these strips to the bottom of the box with sticky tape. Rub the top and sides of the box with the felt or comb your hair and wave the comb close to the box. The strips should stand up and perform an exotic dance. Try it to music!

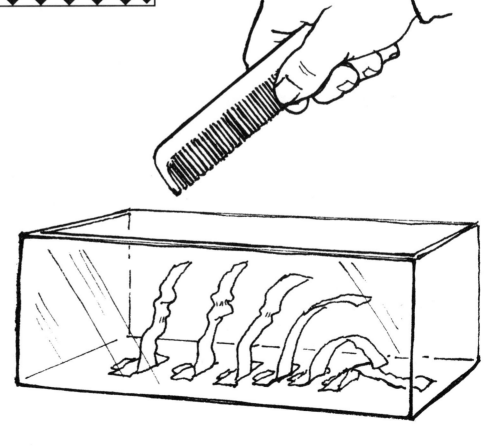

THE DIY ELECTROSCOPE

Electroscopes are the traditional instruments used for measuring electrical charge. They normally contain gold-leaf. This one is easy and cheap to make, using thin aluminium foil.

You will need:

A transparent plastic jar *or* a glass jam jar with its lid

A 3 in (7.5 cm) nail

A chewing gum wrapper (Wrigley's type)

White spirit

Cotton thread

Sticky tape

A plastic comb

Comb your hair and rub the teeth of the comb across the head of the nail. The sticky tape and cotton thread will act as conductors of electricity and take the charge to the pieces of foil. Each piece gets the same charge – so they fly apart.

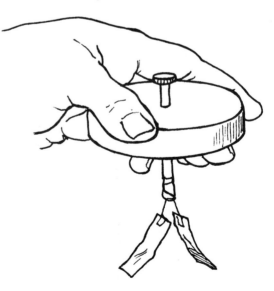

What to do:

Soak the chewing gum wrapper in white spirit for a few minutes.

Knock the nail through the lid of the jar so it's about half way in. Fix in place if necessary. Remove the backing paper from the foil wrapper carefully and cut the foil in half. Fix each piece to either end of a short length of cotton thread using sticky tape. Take the lid off the jar and attach the middle of the thread to the bottom of the nail, again using sticky tape. Put the lid back on the jar.

If you make the electroscope from a plastic jar, it should keep the charge for hours.

CURIOUS CONDUCTORS

All sorts of unexpected things conduct electricity.

You will need:

A (DIY) electroscope

A plastic comb

A match

What to do:

Comb your hair and use the comb to charge the electroscope.

Bring the unstruck match close to the nail of the electroscope. Watch the pieces of aluminium foil – nothing should happen.

Strike the match and bring it slowly towards the nail. The leaves of foil in the charged electroscope should come together.

What does this tell you about flames?

RUBBER BAND MOTOR

Most materials expand when they are heated. However, rubber contracts.

You will need:

Corrugated cardboard from a box

Elastic bands

Thin metal and wood to make a base

A small piece of plastic tube

A sewing needle

A lamp

What to do:

Make two identical rings of corrugated cardboard about 6 in (15 cm) in diameter, and glue them together with the corrugations of each at right angles to one another. Cut a short length of plastic tube about 1 or 1½ in (2.5–4 cm) long. Also make two stiff cardboard washers about 1 in (2.5 cm) in diameter.

Make a support using a block of wood and two strips of aluminium. The aluminium strips must have small dents in them to support the ends of the needle. These are best made with a sharp panel pin.

Put the needle, the cardboard washers and the plastic tube together to form an axle.

For the next bit you will need three hands. Hook an elastic band round one end of the needle, over the cardboard ring and round the other end of the needle. Do the same with the other elastic band. Spread out the bands to form the spokes of a wheel. When you have the wheel fitted into its supports, fiddle about with the elastic until everything is nice and evenly balanced. The wheel must run freely.

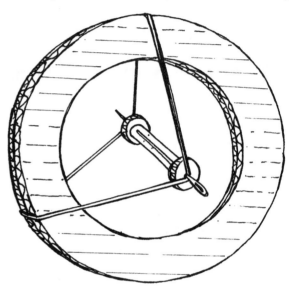

Place a lamp close to one side of the wheel. (You may have to shade the other half.) The wheel should start to revolve. As the rubber bands heat up they contract and pull the wheel off-centre. It over-balances and revolves.

CARBON DIOXIDE
CARBON DIOXIDE

This is a good one to try out at a party of drunken scientists.

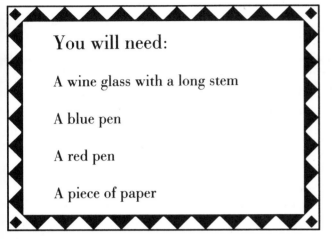

You will need:

A wine glass with a long stem

A blue pen

A red pen

A piece of paper

What to do:

How do you explain that when white light passes through a glass prism it splits into different colours, and when coloured light passes through glass it too is affected to an extent which depends on its colour.

Write the words CARBON DIOXIDE on a piece of paper, CARBON in red and DIOXIDE in blue. Put the paper close behind the stem of the glass and look at the words through the stem. The red letters turn upside down but the blue ones don't.

Why does the glass affect red letters but not blue ones?

It's nothing to do with the words being red or blue. Both words are turned upside down, but because all the letters in DIOXIDE are symmetrical about a horizontal line the word looks the same either way up.

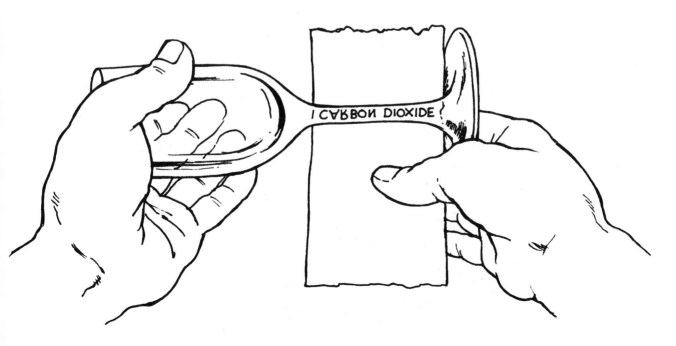

IT'S A GAS!

Carbon dioxide is the easiest gas to make at home and it has some unusual properties.

You will need:

Clear vinegar

Bicarbonate of soda

A tall glass

A lighted candle

What to do:

Put a teaspoonful of bicarbonate into the glass and add a small amount of water. Pour some vinegar into the glass as well. The acidic vinegar reacts with the bicarbonate to produce carbon dioxide, the gas in fizzy drinks. Let the reaction continue for about half a minute then carefully pour the gas, but not the liquid, over the candle flame. The flame will flicker and go out.

Carbon dioxide is a very heavy gas and as it's made it pushes the air away and in this case out of the glass. It's often used for putting out fires because it gets to the root of the flames quickly and starves them of oxygen.

A SOLID GAS

In the days when ice-cream was brought round by horse and cart, everything was kept cold by a strange solid substance called 'dry ice'. Quite often the ice-cream man would let you have a piece which you could use for experiments. Nowadays it's quite difficult to get. However, if you can get some, try this.

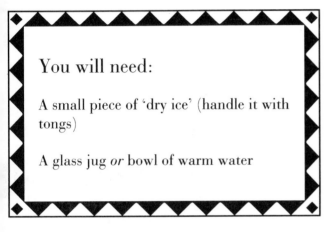

You will need:

A small piece of 'dry ice' (handle it with tongs)

A glass jug *or* bowl of warm water

What to do:

Drop the piece of dry ice into the water. The water will appear to steam and boil, producing huge amounts of vapour.

Dry ice is not ice at all but is solid carbon dioxide manufactured by cooling and compressing the gas. It's a very odd substance because the dry ice turns directly from a solid to a gas (without melting).

The machines which produce mist and fog for stage and television work by dunking dry ice into boiling water.

ANAMORPHIC PHOTOGRAPHS

There are many examples of anamorphic drawings and paintings made to be viewed via cylindrical mirrors (see page 28). This is a way to make photographic versions.

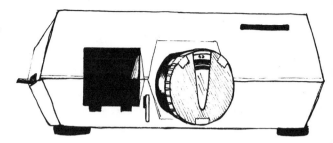

You will need:

A slide projector and slide

A piece of A4 paper and a piece of black card

A cylindrical mirror *or* polished aluminium soft drinks-can (see page 28)

A camera

A tripod *or* the edge of a table

Various books and telephone directories

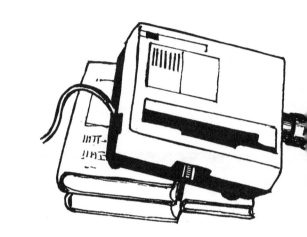

What to do:

Prop the slide projector so that it slopes downwards at about 45° and projects an ordinary colour slide on to the cylindrical mirror or drinks can. The reflection of the slide should fall on to a piece of white A4 paper.

The picture on the paper is obviously very distorted and it's not possible to focus the projector across the entire image. You can improve the focus by putting a 'stop' on the projector lens. A 'stop' is just a device for restricting the amount of light and using a smaller part of the lens. (You may have noticed with an ordinary camera that if you use a 'small stop' or 'lens aperture' the distance over which the lens can keep the picture in focus increases.)

Make a slit in the black card (as shown in the diagram) and attach this to the front of the lens with bits of sticky tape.

The part of the image close to the base of the mirror or can is much brighter than the bit furthest away. The sloping sides of the slit help to even out the brightness.

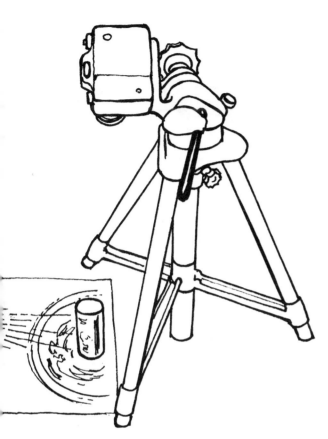

SQUARE ILLUSION

Our eyes and brains are easily deceived – remember 'take nobody's words for it' – especially if they've seen it with their own eyes!

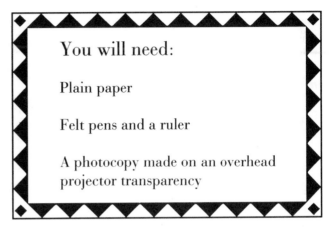

You will need:

Plain paper

Felt pens and a ruler

A photocopy made on an overhead projector transparency

What to do:

Draw a set of radial lines on a piece of paper using a black felt-tip pen. On another piece of paper draw a square with good thick sides. Have the square photocopied on to an overhead projector transparency or draw it directly on one.

Place the square over the centre of the radial lines and then move it side to side. The square will appear to change shape and no longer be square.

Try it with a circle as well.

When everything is ready take a photograph of the anamorphic image from directly over the mirror. Use colour negative film and have an enlargement made to about the same size as the original image. Stand the cylindrical mirror or can in place and view your anamorphic photo.

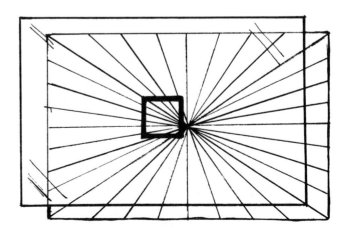

DIY CHAMBER OF HORRORS

Certain parts of the human face are more important for recognition than others. This experiment illustrates which ones they are.

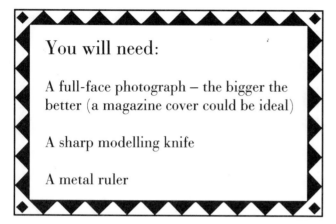

You will need:

A full-face photograph – the bigger the better (a magazine cover could be ideal)

A sharp modelling knife

A metal ruler

What to do:

Using the ruler and knife, carefully cut out the piece of photograph showing only the eyes and another showing only the mouth. Put the pieces to one side.

Turn the rest of the photo upside-down and put the eyes and mouth back in position. Fix them in place using sticky tape on the back of the photo. The eyes and mouth should now be upside-down with respect to the rest of the face. Look at the photo, it should look like a normal face upside-down. Now turn it the right way up – horrors!

Try this with a picture of your favourite politician.

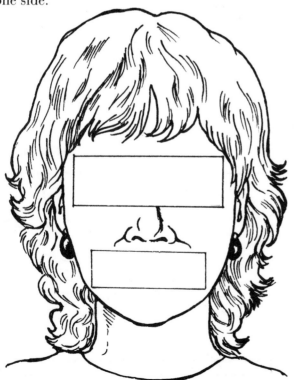

A CORKER OF A PROBLEM

A scientific puzzle to try at parties.

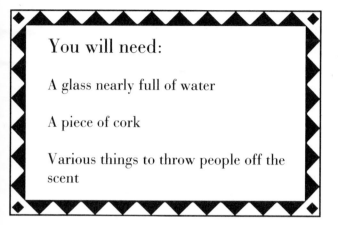

You will need:

A glass nearly full of water

A piece of cork

Various things to throw people off the scent

What to do:

Challenge your friends to make the cork float in the centre of the water in the glass.

The cork refuses to stay in the centre and always floats to the edge of the glass because of surface tension pulling it unevenly. When everyone gives up, show them how it's done.

Pour more water into the glass until the surface of the water is just about level with the rim of the glass. If you are careful the glass will not overflow, surface tension will hold the water in place. The cork will float to the centre because this is now the highest point of the surface of the water.

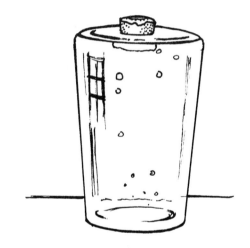

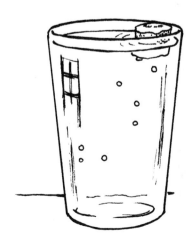

A WHITE CARNATION TURNS RED WHEN WATERED

A bit of chemical magic.

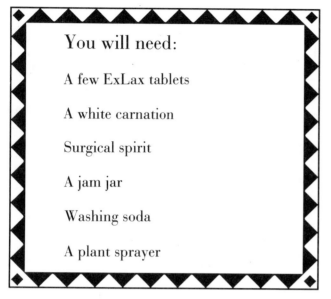

You will need:

A few ExLax tablets

A white carnation

Surgical spirit

A jam jar

Washing soda

A plant sprayer

What to do:

Crush up the ExLax tablets and mix them with surgical spirit in a jam jar. Leave it for a few minutes.

Dip a white carnation into the surgical spirit, remove it and let it dry. Then put it into a vase with some other untreated flowers.

Stand with the vase in front of your audience and, at a suitable moment, spray the flowers with water in which you've dissolved some washing soda crystals. Straightaway the treated flower will go red but the others will stay white.

ExLax tablets contain phenolphthalein (look on the packet) which dissolves in the surgical spirit to produce a colourless solution. When you add an alkali, such as washing soda, to this solution it turns pink or red. Phenolphthalein is one of a group of substances called indicators which can be used to detect alkalis and acids. The red cabbage water used in another experiment is also a good indicator (see page 27).

You can also use phenolphthalein for the old water to wine trick – but don't drink the wine!

MAKING RAINBOWS

Recreate Sir Isaac Newton's world famous experiment.

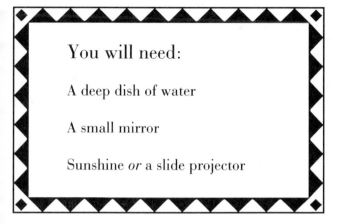

You will need:

A deep dish of water

A small mirror

Sunshine *or* a slide projector

Sir Isaac Newton shone sunlight through a glass prism and produced the colours of the rainbow on the wall of his room. Glass prisms are expensive but you can make your own prism by using a dish of water and a mirror.

What to do:

Make the prism by setting the mirror in the dish at an angle of about 30°. You may need to fix the lower edge of the mirror to the dish with sticky tape. Fill the dish with tap-water.

Allow a strong beam of sunshine or the light from a slide projector to fall across the mirror. Look around the room for a reflection of the white light. Not far away there should also be a bright rainbow. Move the mirror and the dish until the rainbow is in a comfortable place to view it, either on a white ceiling or on a 'screen' of white paper. You will see that white light is made of a mixture of colours.

Transparent things, like water and glass, bend light. Put a pencil in a glass of water and see how it looks.

The amount that the light bends depends on its colour. The white light going into your water prism comes out as a rainbow because the various colours within it are bent or 'refracted' different amounts. You are seeing them spread out on your 'screen'.

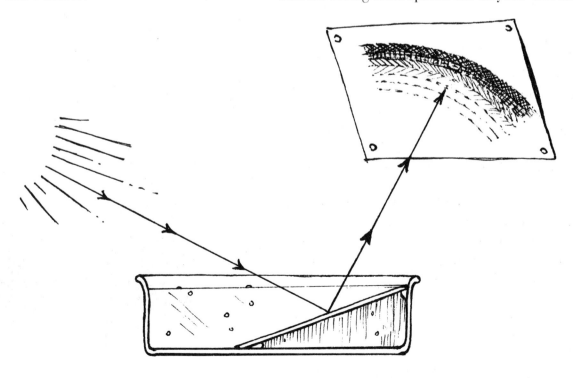

BUBBLES

SEARCH FOR THE PERFECT SOLUTION

Everyone likes blowing bubbles, it's just that we all want to blow them bigger and better than everyone else – here's how.

You will need:

A gadget for blowing big bubbles

Bubblebath for babies

Gelozone thickening agent (available from health food shops)

Glycerine (from a chemist)

A suitable container for the solution

What to do:

Most recipes for bubble solution use washing-up liquid, water and glycerine to produce reasonable bubbles. Recently a new toy came out in the USA which can produce huge long-lasting bubbles, but only if you use particular brands of washing-up liquid like Joy or Dawn. At the moment neither of these are available in Britain. The bubbles produced are brilliant; and they burst in 'slow motion'.

This is a clue. The bubbles must be held together by long molecules. Books about bubbles always emphasise how good the soap must be. So it seems that good soap mixed with long molecules could be the answer. The joy is that this is true.

Baby bubblebath is usually very high quality and free from detergents. Gelozone contains guar gum which provides the long molecules.

Make up some gelozone according to the instructions on the packet. Mix together ½ cup (100 ml) of bubblebath, 1 cup of made-up gelozone, 2 cups (400 ml) of tap-water and if possible ¼ cup (50 ml) of glycerine. You can do this while the gelozone is still warm. The resulting luke-warm solution is at the ideal temperature.

Dip your amazing bubble gadget into the solution and try to blow the biggest bubble in the world. I suggest you do it outside.

Try varying the quantities or leaving out a component from the mixture. Maybe you can find a substitute for the gelozone – what else contains long molecules?

BIG-BUBBLE GADGETS

Any gadget for making large bubbles has to be able to hold lots of bubble solution. Simple wire frames made from coat-hangers are OK but a strip of cloth or string will hold more bubble solution.

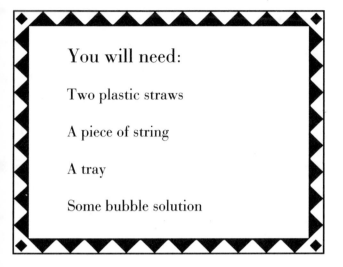

You will need:

Two plastic straws

A piece of string

A tray

Some bubble solution

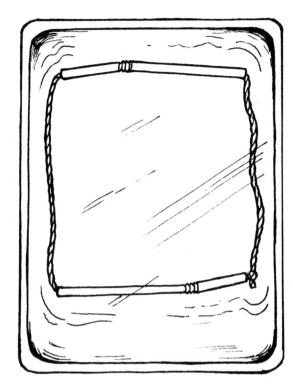

What to do:

Thread about 3 ft (1 m) of string through the straws and tie the ends together to make a rectangular frame.

Dip the straws and string into a shallow tray of bubble solution, at the same time making sure your hands are good and soapy. Pick up the straws with your finger tips, keeping them close together. Stand up and open the frame but keep the strings fairly loose. Allow the breeze to catch the soap-film and blow a bubble. When the bubble has formed close it off by bringing the straws together.

If there's no wind, pull the frame upwards to form the bubble. It takes a bit of practice to close the bubbles off without bursting them.

But it's well worth the effort.

The best bubble gadget around is a commercial toy consisting of a plastic rod about 2 ft 6 in (75 cm)

long. A short and a long strip of lacy material are fixed to one end of the rod. The other ends of the material are fixed to a ring which slides on the rod. The free end of the rod is held and the ring is pushed up to the fixed end. The material is dipped into the bubble solution and well soaked.

After the gadget is removed from the solution, the ring is pulled along to open up the loops of material and produce a film of the bubble solution. If there is a slight breeze it will catch the film and blow a long bubble which can be nipped off by moving the ring and closing the loop. When there is no wind the bubble is produced by waving the rod and walking along. Some people claim to have made bubbles 15 feet long. The huge bubbles are a wonderful sight as they wibble and wobble through the air.

SLAP-STICK SLIP STICK

The friction between two objects is lower when they are sliding over each other than when they are stationary.

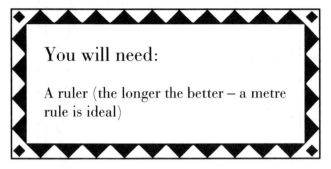

You will need:

A ruler (the longer the better – a metre rule is ideal)

What to do:

Hold the metre rule at each end, just resting it on your forefingers. Keep one hand stationary and move the other towards it. At what point do you think the ruler will overbalance?

You'll find that your fingers always finish together with the ruler perfectly balanced. Start with your fingers in any position under the ruler. The result will always be the same.

As one finger starts to slide under the ruler, it takes more of the weight and the friction between that finger and the ruler increases. It gets to a point where that friction is higher than the static friction at the stationary finger. Then the ruler starts to slide over the second finger. As soon as the ruler slips over the second finger the friction drops. The friction at the first finger is now quite a bit higher and sliding stops at that finger. This to and fro continues until both fingers meet under the centre of the ruler.

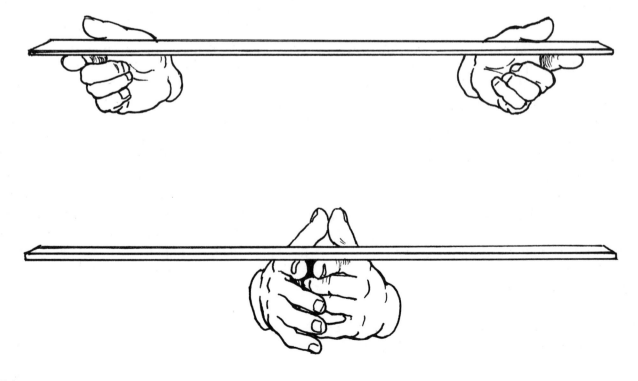

MAGIC MOTOR

A simple propeller which rotates every time you pick it up.

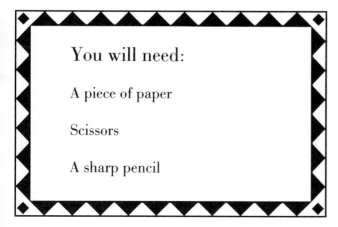

You will need:

A piece of paper

Scissors

A sharp pencil

What to do:

Trace the shape shown in the diagram on to a sheet of ordinary writing paper. (Or just make a photocopy.) Cut out the shape and fold it in half at the angle shown. Open up the propeller blades and balance it on the point of a sharp pencil.

If you hold the pencil upright the propeller should spin. Your hand warms the air, this warm air rises past the blades and makes them turn.

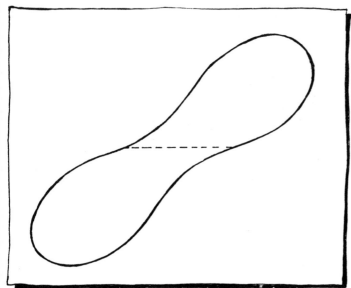

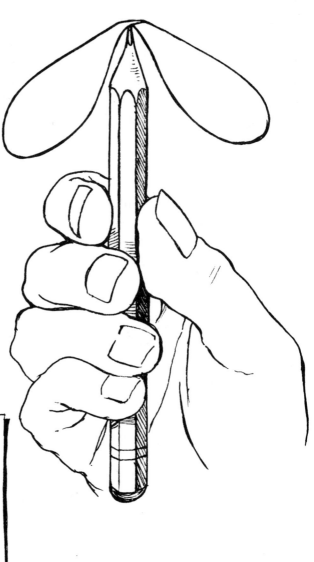

STICKING PINS IN BAGS OF WATER

Modern plastic materials have amazing properties – here's an amusing trick.

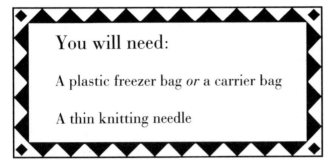

You will need:

A plastic freezer bag *or* a carrier bag

A thin knitting needle

What to do:

Three-quarters fill the bag with tap-water. Hold it fairly loosely by the neck with one hand, and briskly stab the needle through and out the other side with the other hand. All things being well you will lose little or no water – provided the needle stays in place. The pressure of the water inside the bag forces the plastic film to form a watertight seal around the needle.

Practise over the sink and then try it over a friend's bed – with them watching of course!

Hold the bag over the sink and remove the needle. Water will squirt out. Now try and put the needle back through the original holes. Can you get them to seal up again?

Empty the water from the bag and examine the holes. They should be neat round holes with stretched edges which formed the seal.

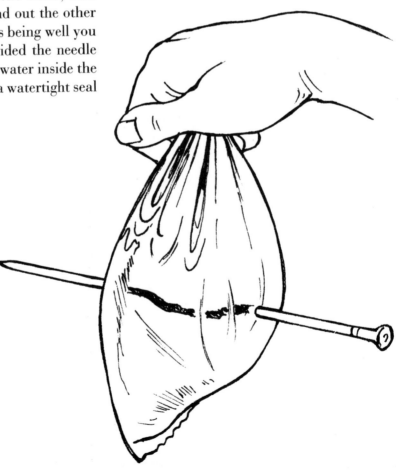

THE YOGHURT POT MEMORY TEST

Yoghurt pots can be made to remember what they were before they became yoghurt pots.

You will need:

A yoghurt pot (well – what else?)

A pressure cooker *or* a pan of boiling water *or* an oven

What to do:

Put about 1 in (2.5 cm) of water in the pressure cooker, drop in a clean yoghurt pot and close the lid. Boil for a few minutes using the 15 lb (7 kg) pressure weight. Cool the cooker and take out the yoghurt pot. It should have turned into a flat disc a few inches in diameter. The writing and pictures on the original pot will be distorted and spread round the edge.

You have an anamorphic picture of a yoghurt pot.

If you don't have a pressure cooker, boil the yoghurt pot vigorously in a covered pan of water. It will take much longer to collapse and may not finish up completely flat, as the material of the yoghurt pot seems to need more than 100°C to relax back to its original form.

Put lots of salt in the pan of water to increase its boiling point.

Another way to collapse the yoghurt pot is to heat an oven to about 230°F (110°C). Stand the pot on an aluminium foil pie-dish and leave it in the oven for a few minutes. If your oven has a glass door you can watch the pot shrink.

Plastic materials come in two kinds: thermoplastics, which soften when reheated, and thermosetting which resist further heat treatment.

You can shrink one of your own designs if you get a plain white pot. Draw your design with a fibre-tip pen (the silver pens are especially good), then shrink the pot.

If you stand a little cylindrical mirror (see page 28) in the centre of the disc, you should see your original design.

MAKE YOUR OWN MICROSCOPE

Not many people own microscopes but binoculars are fairly common. These are easily converted into a good microscope without causing any damage.

You will need:

A pair of binoculars

A magnifying glass

Blu-Tack *or* sticky tape

Various bits of wood and screws

Black paper

There are several simple ways to provide fine focusing. Put the object you want to inspect on to a piece of foam rubber and squash the foam until the image is in focus. Of course it only stays in focus while you hold it in place. But it's perfect for looking at 3-dimensional objects.

You can get a bit more control by making some wooden wedges or using the edge of a ruler. Put the object on a flat dish or plate and push the wedges underneath. Moving them will change the distance between the microscope and the object.

What to do:

The conversion is very straightforward. The magnifying glass has to be attached to the objective end (not the eyepiece end) of one side of the binoculars. You can do this with sticky tape or Blu-Tack, so the conversion is temporary.

The focusing knob on the binoculars will not focus the 'microscope'. Try it out by looking at some newsprint. You have to move the whole thing nearer or further from the object. Once you get the idea, it's worth making some sort of stand for the new microscope. This is probably best made from wood.

If you use a powerful magnifying glass, you'll get more magnification but it will be difficult to focus clearly across the whole image. There's an easy solution to this problem, you have to reduce the effective size of the magnifying glass. Cut out a disc of black paper the size of the lens and make a small hole, about ¼ in (1 cm) in diameter, in the centre.

Fix this disc to the magnifying glass. Now, when you look through the microscope, the focus should be much improved (though the image will be dimmer). Experiment with various-sized holes until you are happy with the results. You may have to shine more light on the object.

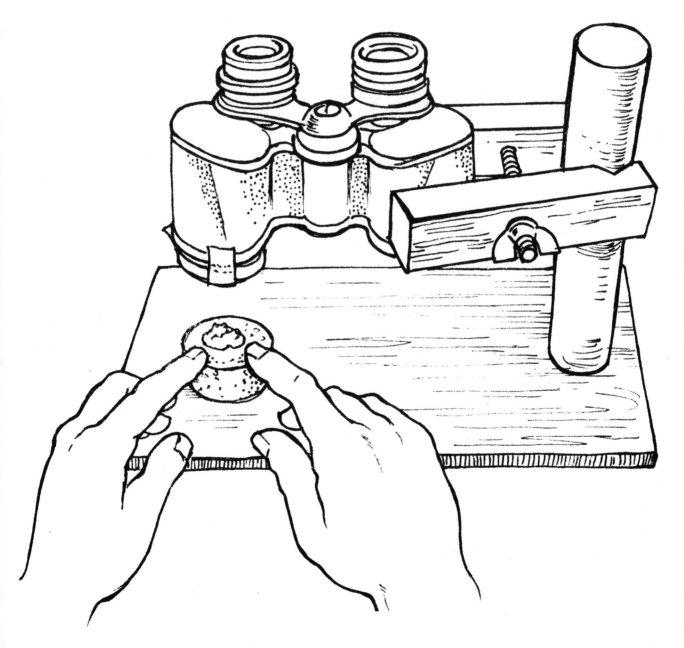

THE LOOKING-GLASS WORLD

AN INSIDER'S VIEW

We often look down at the surface of the water. In this experiment you can see it from underneath – it's a whole new world.

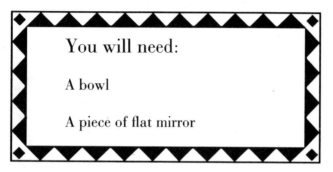

You will need:

A bowl

A piece of flat mirror

What to do:

Fix the mirror into the bowl at an angle of about 45°. You may have to stick the bottom edge down with tape. Fill the bowl with water and let it settle. Look straight down at the mirror and gently put your finger tip in the water. You should see your finger appear as if from nowhere with no sign of the rest of your hand, or the outside world.

As you lower your finger into the water yet another reflection will appear, this time pointing in the opposite direction.

Your finger seems to come through a liquid mirror, like mercury, and even to repel the water.

Make some bubbles on the surface and look at them from 'inside'. The 'inside' surface of the water acts like a perfect but liquid mirror.

Good for special effects maybe?

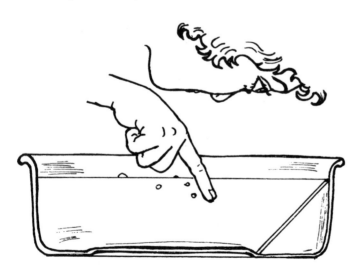

A VORTEX IN YOUR HANDS

When a vortex occurs in nature, such as a whirlpool or a tornado, it's very difficult to see what's really happening. This experiment puts one in your hands.

You will need:

Two clear plastic lemonade *or* mineral water bottles with their tops

Adhesive such as Bostik *or* Araldite

A drill *or* some method of making a hole

What to do:

Stick two bottle tops back to back as accurately as you can with a good adhesive like Araldite. When this has set make a hole about 5/16 in (8 mm) in diameter in the centre of the bottle tops. A drilled hole is best.

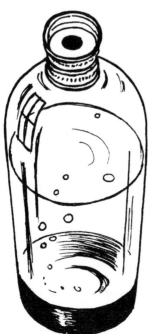

Three-quarter fill one bottle with water, put on the special double cap and screw the empty bottle on top. Now turn the device upside-down. Very little water should run through from the top bottle to the bottom one.

Give the whole thing a gentle swirl to start the water rotating. If you do it enough a whirlpool or vortex will form as the water pours through.

Turn it over and try it again.

CRYSTAL CAVES

Growing single crystals takes a long time, and what can you do with them afterwards? This experiment is quicker and more fun.

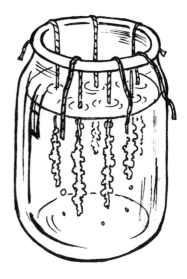

You will need:

A small waterproof box

Potash alum (from a chemist or shop specialising in dyeing)

Cotton thread *or* pieces of flannel

Sticky tape *or* plasters

Small bits of coloured filter *or* sweet papers

A small lens with a focal length suitable for the size of the box

When you've seen the effect, make several holes in the sides and one in the lid of your box, then cover them with sticky tape or plasters to make the box watertight again.

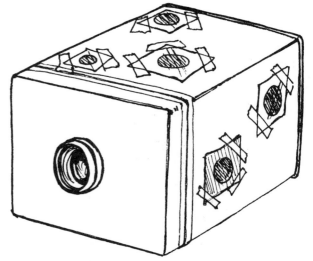

What to do:

Make up a solution by dissolving as much potash alum as you can in hot water. Pour it into a jam jar and suspend several pieces of cotton thread in it. As the solution cools crystals of alum will start to grow round the threads.

Arrange cotton threads or pieces of flannel in the box and then pour the hot alum solution through the hole in the lid.

Leave to cool and crystallise, and then drain out any spare solution and allow everything to dry. Remove the sticky tapes and cover the holes with coloured filters or sweet papers. Fix a small lens to the hole in the lid and view your crystal cave.

A CAFÉ TABLE EXPERIMENT

A humble cup of tea shows an amazing effect.

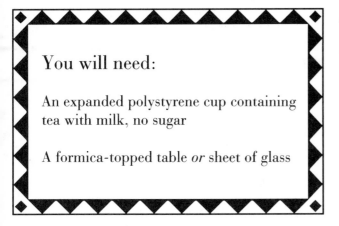

You will need:

An expanded polystyrene cup containing tea with milk, no sugar

A formica-topped table *or* sheet of glass

What to do:

The table needs to be squeaky clean.

Push a cup of tea slowly across the surface of the table. If you get the speed right the surface of the tea will form a stable pattern or standing wave. Now press the cup slightly harder or faster across the table to get more energy into the tea. Globules of the liquid will start to jump out of the cup all over your hand and the table.

Judge the speed or pressure a little more carefully and you should be able to produce the globules without showering your hand or the table. Look what happens when they fall back on to the surface of the tea. Rather than just rejoining the liquid they roam around the surface for quite a few seconds. (We've had them stay on top for as long as 6 seconds.)

Try the experiment with various amounts of milk, different strengths of tea and even a little sugar. The little spheres seem to be separated from the surface of the tea by a thin layer of air.

RHAPSODY IN GLUE

Glue doesn't come from tubes in the same way that milk doesn't come from bottles! But did you know that glue comes from cows? Here's how to make it.

You will need:

Some skimmed milk

Vinegar (clear is best)

Bicarbonate of soda *or* washing soda

A pan

What to do:

Warm a pint of skimmed milk and add at least 2 tablespoons of vinegar stirring all the time. A sticky white substance will separate from the milk and leave behind a thin yellowish liquid. The white stuff is casein, one of the proteins in milk.

Milk is normally nearly neutral pH but adding vinegar makes it quite acidic and allows the casein molecules to clump together or curdle. Lots of salt will have a similar effect.

To make the casein into a glue you have to dissolve it in an alkali. Strain off the casein lumps from the liquid and mix them with 2 or 3 tablespoons of warm water, and about 6 teaspoons of bicarbonate of soda or a solution of washing soda. Mash this well with a fork and force it through a sieve.

The casein dissolves quite slowly in the alkali so leave it in a covered pot for about a day. You should end up with a milky liquid which will work as a glue.

Casein is also used commercially in the manufacture of paper and plastics.

Perhaps you can get it to set in small moulds and make plastic objects from milk.

PAPER MARBLING

Look inside the covers of old books and you will often see exotically decorated endpapers. It's not too difficult to produce similar patterns (called marbling) at home by floating paint on the surface of another liquid.

You will need:

Either non-fungicidal wallpaper paste (made up to ½ strength), gelatine or gelozone (see page 52)

Oil paints (in tubes)

Linseed oil

A shallow waterproof tray

A knitting needle *or* special comb (see page 67)

Various pots, sticks and jam jars

White spirit

Vinegar

An apron!

What to do:

The first thing to prepare is the base liquid on which you will float the colours. Try a little experiment. Mix a small amount of oil paint with white spirit until it's quite thin and runny. Drop small amounts of this on to water and watch it spread out. It's this spreading out that needs to be controlled when you make marbled paper. The base liquid needs to be thicker or more viscous than plain water.

Mix 1 oz (28 g) or about 2 dessertspoons of gelatine with 1 pint (½ litre) of hot water. When it's completely dissolved add about the same amount of cold water. The gelatine should remain liquid when it's cool.

Pick the colours you wish to use and in separate pots mix about 1 in (2.5 cm) of each with enough white spirit to make them runny. Spatter small amounts of these colours on the surface of the gelatine solution. They should not spread out too much (1 in wide is about right). Stir the floating colours with a knitting needle or a special comb (see over) to produce exotic patterns.

Lay a sheet of paper carefully over the pattern. Hold the paper at either end so it sags in the middle, and let this down on to the colours. Then, gently let the edges down, and prevent any bubbles from forming underneath. Lift off the paper and let it drain for a few seconds and then lay it flat on old newspaper to dry.

If the colours seem to be breaking up, mix a few drops of linseed oil with the paint and try again. You could also add one or two teaspoons of vinegar to the base solution.

Experiment by adding a drop of washing-up liquid or baby bubblebath to the paint. This noticeably affects the way the colours behave on the surface of the gelatine solution.

This kind of marbling depends on the fact that oil paints don't mix with water; and that surface tension can be altered by additives, like white spirit, linseed oil or washing-up liquid. The thickness or viscosity of the base solution is also important. For good results you'll have to experiment with the thickness of the base solution and the dilution of the oil paints.

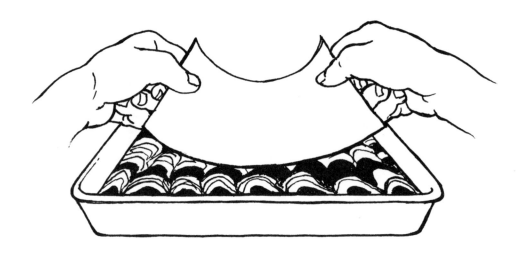

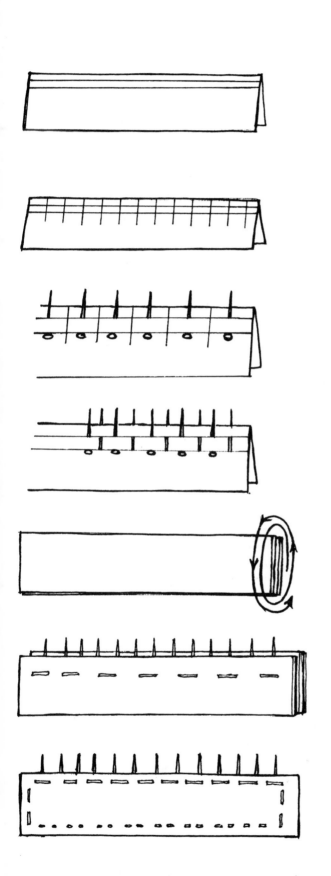

To make a special marbling comb you will need paper, card, a ruler, a pencil, some scissors, some pins and a stapler.

Cut a strip of paper the same length as the paper you want to marble and about 2½ in (6.5 cm) wide. Fold the strip in half lengthwise, and draw lines ¼ in (0.5 cm) and ½ in (1 cm) from the folded edge on both sides of the strip.

Now make marks every ½ in (1 cm) along the lines you have already drawn (as shown).

Insert pins through the double thickness of paper where the lines intersect at each alternate point.

Turn the strip over and insert pins on the remaining marks.

Cut four strips of card (from a box of a cereal packet) the same size as your marked strip, and use double. Cover each pair tightly with clingfilm. Wrap the clingfilm round twice for strength.

Staple your paper strip to one of the card strips making sure that the paper is taut, then sandwich it with the other card strip. Staple close to all sides (as shown).

Turn the comb over and lightly hammer the staples down. Your comb should be fairly strong.

PENCILS CONDUCT ELECTRICITY

'Lead' pencils are really made from graphite which is a conductor of electricity.

You will need:

A lead pencil

A modelling knife *or* penknife

A torch bulb

Thin wires

A battery

What to do:

Carefully split the pencil along its length using a knife. (Always cut away from yourself.) The lead should stay in one half of the wooden casing. Wrap wires round the ends of the lead and connect to the battery and bulb.

The bulb will be dimmer when connected via the lead than when connected directly to the battery. By sliding one of the wires along the lead you can vary the resistance and the brightness of the bulb. Electrical resistance is measured in ohms.

The hardness of a lead pencil depends on how much clay is mixed with the graphite. The more clay the harder the pencil and the higher the resistance. 4H pencils are about 20 ohms, HB pencils about 2 ohms.

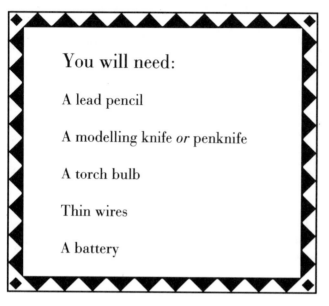

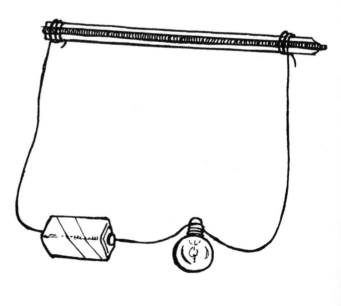

COPPER AND SILVER PLATING NON-CONDUCTORS

If you dip an iron nail into a solution of copper sulphate the nail will become plated in copper. Simple. It is much more interesting to plate something which is not metal. Here's how.

You will need:

A leaf *or* a toy *or* anything you want to plate

Copper sulphate (from a chemist, a garden centre, a chemistry set or a crystal garden kit)

A watertight plastic box *or* bowl

A piece of copper (plumber's suppliers)

A 6-volt battery *or* low-voltage DC supply

Various wires

Varnish

Methylated spirits *or* thinners

Graphite powder (from a model shop) *or* copper powder

A resistor *or* a 4H pencil *or* 6-volt 0.1-amp torch bulb

Rubber gloves (wear these at all times during this experiment)

What to do:

Make a solution of copper sulphate equivalent to 2 oz in a pint of water (or 100 g in a litre). Dip a steel nail or screw into the solution for a few seconds. It will plate with a pink layer of copper. Rub the coating with your finger and it will come off easily because it's only a few atoms thick.

To produce a thicker coat, you have to pass an electric current through the solution. You should connect the object you plan to copper-plate to the negative (or cathode) of a battery and a small piece of copper to the positive.

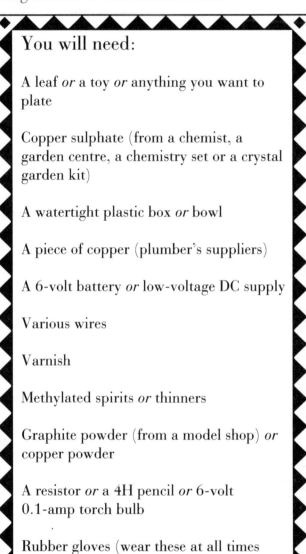

To copper-plate something which doesn't conduct electricity, you first have to turn it into a conductor. The simplest way to do this is to coat the object in graphite powder. Mix up some varnish (or polyurethane) with methylated spirits (or thinners) and paint this on to the object in question, making sure every bit is covered. Put some powdered graphite in a plastic box. (An ice-cream container is ideal.) When the object is tacky-dry, put it in the box and shake it up with the graphite until well coated.

It's important that the electric current used for plating is kept quite low. (If you have access to an ammeter the current density should be no more than .04 amps per square inch (or .01 amps per square centimetre) of surface to be plated. Guess or calculate the area of the object to work out the ideal current.)

You have to include some resistance in the electrical circuit to prevent the current getting too high. A 50 ohms resistor is about right, but if you don't have one you can use a pencil. (See page 68.) A 4H pencil has a resistance of about 20 ohms but you can use two connected in a series.

Put the copper sulphate solution into a deep bowl and connect up the circuit. Attach the object to be plated securely to the negative of the battery and connect the piece of copper pipe with a wire through the resistor to the positive. A good contact is important. Immerse the object and the copper pipe in the solution, making sure they don't touch. Switch on the power and leave for several days. Check regularly.

You may find it helps to add a teaspoon of vinegar to the copper sulphate solution. The acid in the vinegar makes the solution a better conductor. Keep the plating going until you have produced a good thick coat of copper.

You can silver-plate the object by rubbing the copper with one of the commercial silver-plating compounds such as 'The Silver Solution'. Or, if you prefer to keep the copper pink, coat it in clear varnish diluted with methylated spirits.

Another way to produce a graphite coating is to mix the graphite with a small amount of strong hair gel and paint this thick mixture on to the object. A bit more care is needed to get an even coat of graphite but it works quite well.

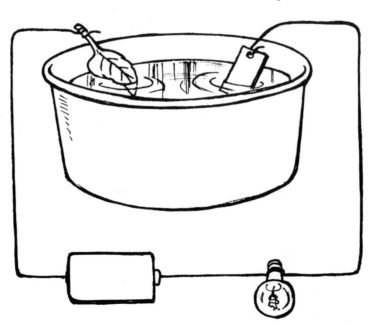

PHOTOGRAPHING A TV PICTURE

Many people now have videos of their wedding or children and would like to produce photos from them to give to grandparents or other relatives – here's how.

> ## You will need:
>
> A tv
>
> A suitable camera
>
> A video recorder
>
> A tripod *or* a pile of books

What to do:

There are a couple of points worth making at the start. Firstly, not all still cameras are suitable. Secondly, a tv picture is deceptive, it looks better than it really is. Don't expect the results to be as good as original photographs.

Any tv is suitable but large-screen models are better if only because it's easier to focus on the screen with the camera.

A tv picture changes 50 times a second. This is the cause of most of the problems of trying to photograph the screen as a camera shutter speed faster than $\frac{1}{50}$ second will capture only part of a whole picture. So for a camera to be suitable for this experiment it should have an adjustable shutter speed. Though many modern cameras have fully automatic exposure and the shutter speed changes they don't allow the user to alter it directly. However, these cameras may still be suitable. Older cameras and more expensive modern ones have a variable shutter speed which you can set manually. These are the best.

A tv picture is made up of 625 lines, with half this number being transmitted every ⅟₅₀ second. Lines 1, 3, 5, 7 and so on are sent, then ⅟₅₀ second later lines 2, 4, 6 etc. will follow. This method of transmission is called 'interlacing' and produces a full-definition picture every ⅟₂₅ second. So to photograph this you will need a shutter speed longer than ⅟₂₅ second (usually ⅟₁₅ or ⅛).

100 ASA colour film is suitable for a normal tv set. If you have one of the newer video recorders, then you can set it to hold a good still frame and you will have no problems with the slow shutter speed.

Fully automatic cameras present much more of a challenge, but many can still be made to work. Look in the instruction book to see whether the camera automatically alters its shutter speed. It

should state the lowest shutter speed the camera will produce. Otherwise, next time the camera is empty, point it at a dark subject, release the shutter and listen carefully. Now point it at a bright subject, the window or a light, and do the same. Listen for a difference in shutter noise. Does one sound slower or clunkier? If so, the camera does alter its shutter speed. Your problem then is how to get the speed you require. With 100 ASA film loaded the automatic shutter speed will be a little too fast, but there are several ways to slow it down:

If your camera has a backlight setting use this

or set the ASA rating at 50 instead of 100

or turn the tv screen brightness down a little

or put something over the magic eye exposure meter to make it 'think' the subject is darker than it really is. Half cover it with dark tape or stick a piece of greaseproof paper over it.

The picture will be a little bit over-exposed but with most modern films this is not critical. Some cameras have a built-in flash unit which will trigger automatically at low light settings. Of course the flash will spoil the tv picture so you'll have to shield it with a piece of card or thick paper.

When the shutter speed is low the camera needs setting on something solid to keep it level and still. A tripod is ideal for this but a chair and a pile of books will do. Be careful not to position the camera closer than its minimum focusing distance.

DIY AERIAL PHOTOGRAPHY

Most aerial photography is done from an aeroplane or possibly a balloon and is quite expensive. This method uses a kite and you can do it yourself.

You will need:

A large kite (a delta type is recommended)

Wood, string and glue

A camera

A cable release and attachment if necessary

String to operate the shutter

A ball and socket camera mount

The wind

What to do:

A lightweight camera with a motor drive is ideal, but practically any camera can be used. If you don't have a motor-driven camera you just have to bring it down each time to wind the film on.

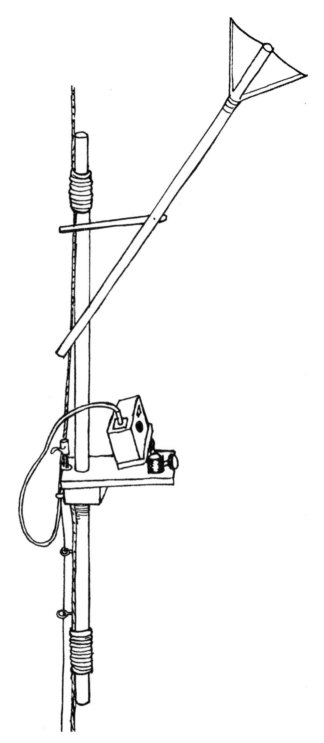

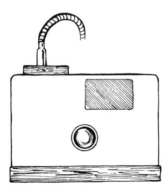

The camera has to be mounted on a rig which is attached to the kite string. The diagram shows the kind of rig which works well. It should be flown about 50 ft (15 m) below the kite. The tail keeps it stable in the air.

Many modern cameras don't have a cable release socket but you can buy attachments which screw to the camera (see bottom right). The ball and socket mount is a useful refinement which allows the camera to be positioned easily.

If you don't want to risk an expensive camera for your first flight, it's easy to adapt one of the very cheap cameras available nowadays. You need to glue two small pieces of wood to the camera, one piece takes the screw for the cable release and the other the screw for the camera mount. If you use one of these cheap cameras, the rig could be simplified by fixing the camera in place with sticky tape instead of using a ball and socket mount.

Thread the two pieces of plastic tube on the kite string before you launch the kite. Fly the kite up to about 50 feet (15 m) before you attach the camera rig to the kite string. The main dowel of the rig is pushed through the plastic tubing and it should be a snug fit to secure it to the kite string. The shutter is released by pulling the other string.

It's important that the shutter string is attached in such a way that it doesn't shake the camera when the string is pulled.

In Britain there are a few flying restrictions for kites: you should not fly them higher than 200 feet (60 m), nor within 5 miles (8 km) of an airfield or near to power lines. Even so, it's still possible to get excellent aerial photographs from about 150 feet (45 m) up.

LIQUID MAGNETS

If you lay a piece of paper on top of a magnet and sprinkle the paper with iron filings, you can see the two-dimensional pattern produced by the magnetic field. This experiment is much more gooey!

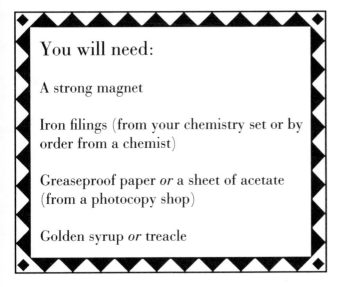

You will need:

A strong magnet

Iron filings (from your chemistry set or by order from a chemist)

Greaseproof paper *or* a sheet of acetate (from a photocopy shop)

Golden syrup *or* treacle

What to do:

In a small pot or jar mix together the iron filings with a small amount of the syrup or treacle. Put some of this goo on to the acetate sheet, slide the magnet underneath and move it about. The iron filings will flow under the pull of the magnet and produce a three-dimensional view of the magnetic field. They usually form miniature mountain ranges.

Try mixing the filings with some clear thick liquid (glycerine or liquid paraffin) in a jam jar and then hang a small magnet from a piece of thread in the liquid. Slowly the filings will move and surround the magnet making a sort of fur coat representing the shape of the magnetic field.

SLIMY YUKKY STUFF

This stuff is wonderful and very messy!

You will need:

PVA glue (stationers, toy shops, DIY shops)

Borax (chemist)

Glycerine (chemist)

Plastic *or* paper cups

A stirrer *or* lolly stick

Rubber gloves if you have skin problems or cuts

A plastic teaspoon

Stir it well. The mixture will stiffen up and form a sort of rubbery stuff. Take it out of the cup and gently knead it by hand. You should be able to make it into a ball which will bounce. If you pull it slowly it stretches, pull it quickly and it will snap.

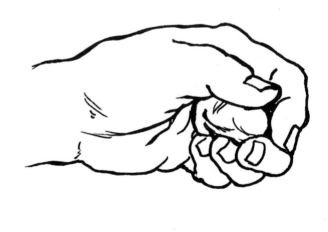

What to do:

You can vary the quantities of ingredients quite a lot but try it this way to start with. Dissolve 1 teaspoon of borax in about half a cup of warm water. Into another cup put two teaspoons of PVA glue and add one teaspoon of the borax solution.

PVA glue contains long chains of molecules which are surrounded by water. The borax makes links form across these chains and this produces a more rigid structure than the glue itself.

You can make your new rubbery stuff less stiff by adding some water to the glue before you mix in the borax. Mix together 2 teaspoons of PVA glue with 4 teaspoons of tap-water, then stir in borax a teaspoon at a time. The crosslinking process is not so rapid and you can actually watch the thickening take place. Add more borax solution until you get a good consistency. This stuff is much more floppy but it will still bounce a bit.

To produce a slimy effect, you need to add a plasticiser to the mixture. Glycerine is just the job for this. Mix 2 teaspoons of PVA with 1 teaspoon of glycerine and then stir in some borax solution a little at a time. The stuff will be much more sticky and you may need to add more glue to get the best result. It's quite manageable but watch the furniture!

Write on a piece of paper with a variety of pens and markers. Press a piece of the stiffest stuff on to the writing and any water-based ink will lift on to it.

WHY IS THE SKY BLUE?

OR WHY IS THE SUNSET ORANGE?

You will need:

A goldfish bowl *or* a rectangular aquarium

Dettol *or* milk

Sunshine *or* a slide projector

A mirror

A piece of white paper

What to do:

Fill the bowl or aquarium with tap-water and shine a strong light through it. Sunshine is best but a bit unmanageable. However, you could use a mirror to direct it where you want it. If you use a slide projector, put in a slide containing a very pale blue filter, and a piece of black paper with a hole in it about the size of a penny. This will be more like sunshine.

Let the light coming out of the bowl fall on to a piece of white paper. Notice the colour of the light.

Add a few drops of Dettol to the water and stir it. The water will become milky and the light will start to turn yellow. Keep adding small quantities of Dettol and the milky water will take on a bluish hue. This will be especially obvious when you look at the beam of light coming from the slide projector.

The light falling on the paper 'screen' will go orange and fade out as the Dettol takes effect and the 'sun' sets. Blue light is scattered from the 'sunbeam' by minute particles in the water. The same thing happens in the earth's atmosphere where molecules of air and fine dust scatter the blue part of sunlight, but allow the yellow or orange parts to get through.

At sunset the light has to travel through more atmosphere so more of its blue light is scattered, and the sun looks orange or red.

The extent to which light is scattered depends on its frequency multiplied by itself four times. Though blue light has a frequency only twice as high as red light it is actually scattered about sixteen times as much.